AF709626

Security glazing: is it all that it's cracked up to be?

A guide to the selection of effective security glazing

Craig Devine and Richard Flint

This work has been funded by BRE Trust. Any views expressed are not necessarily those of BRE Trust. While every effort is made to ensure the accuracy and quality of information and guidance when it is first published, BRE Trust can take no responsibility for the subsequent use of this information, nor for any errors or omissions it may contain.

The mission of BRE Trust is 'Through education and research to promote and support excellence and innovation in the built environment for the benefit of all'. Through its research programmes the Trust aims to achieve:

- a higher quality built environment
- built facilities that offer improved functionality and value for money
- a more efficient and sustainable construction sector, with a higher level of innovative practice.

A further aim of BRE Trust is to stimulate debate on challenges and opportunities in the built environment.

BRE Trust is a company limited by guarantee, registered in England and Wales (no. 3282856) and registered as a charity in England (no. 1092193) and in Scotland (no. SC039320).

Registered Office: Bucknalls Lane, Garston, Watford, Herts WD25 9XX

BRE Trust
Garston, Watford WD25 9XX
Tel: 01923 664743
Email: secretary@bretrust.co.uk
www.bretrust.org.uk

BRE Trust and BRE publications are available from:
www.brebookshop.com
or
IHS BRE Press
Willoughby Road
Bracknell RG12 8FB
Tel: 01344 328038
Fax: 01344 328005
Email: brepress@ihs.com

Printed on paper sourced from responsibly managed forests

FB 55

First published 2013
ISBN 978-1-84806-305-1

The authors gratefully acknowledge BRE Trust for funding this programme of research and the production of this guide.

In addition, the project would not have been possible without the active participation of Hamilton Erskine and, in particular, Mike McNeil for sourcing the glazing materials used to conduct the tests during the research programme, and the following for providing comments during the drafting of this guide:

- Colleagues at BRE Global Ltd
- Association of Chief Police Officers (ACPO)
- British Security Industry Association Ltd
- Centre for Protection of National Infrastructure (CPNI)
- Gary Heward at MFD International Ltd
- RISC Authority Security Group/Fire Protection Association
- Salford City Council

Contents

Executive summary

The BRE Trust-funded research project 'Security glazing: is it all that it's cracked up to be?' followed a spate of burglaries in the West End of London in 2008. The burglaries in London used 'smash-and-grab' attacks to steal expensive goods such as jewellery and designer clothing from window displays. This showed that of all building elements, glazing offers the least resistance to impact and explosive loading. This was further demonstrated in the 2011 series of riots in UK major cities. Additionally, the majority of injuries from explosive events, including terrorist attack, invariably result from flying glass particles. As such, improvements in the design of glazing to resist attack are required.

Standards are critical to the selection of an appropriate product. However, in the case of glazing a number of different standards have been developed based upon arbitrary methods. For example, glazing may be subjected to a simulated manual attack or impacted by a known mass at a known velocity. The results from such tests are not typically comparable with each other, and so it is not immediately obvious which standard is best suited to a particular application.

This publication has been produced from a BRE Trust-funded research project to investigate the performance of glazing in security applications. It outlines the different types of glazing, guides the reader through the maze of applicable standards and offers advice on selecting glazing systems appropriate to their intended use.

1 Introduction

Naturally occurring glass, such as obsidian, has been used since the Stone Age for the production of sharp cutting tools, and the first evidence of man-made glass can be traced back to 3500 BC in Mesopotamia. Glassmaking manuals can be dated back to 650 BC, where instructions on how to make glass are contained in cuneiform tablets.

Early glass was a luxury item and considered as precious as gold. Egyptians created glass using a method called 'core-forming', whereby molten glass was wrapped around a core (made of clay and dung) and then shaped by rolling the glass on a smooth surface. Glass-blowing was discovered by AD 14; this process of making glass has changed very little since then.

The material began to be used for architectural purposes by the Romans, following the discovery of how to make clear glass around AD 100. Cast glass windows then began to appear in the most important buildings in Rome. Glassmaking subsequently spread across Europe and beyond.

The techniques used to produce glass continue to advance even today, resulting in clearer glass with less impurities and imperfections. Glass has been used for a variety of applications, from jewellery to mosaic tiles, from storage containers to stained-glass windows.

Probably the most common type of glass we see today is that used to form the glazing of windows in the buildings we live and work in. Glazing not only allows natural light to permeate into buildings, it must also keep the elements out and help to maintain a comfortable environment within the building. Glazing also needs to provide a host of other performance attributes, including acoustics, fire resistance and security.

This guide was primarily developed to aid the selection of effective security glazing. It will help architects, consultants, police officers and others who advise people on the selection of effective security to determine whether the glazing provides an equivalent level of resistance to forced entry by criminals as that provided by the building elements into which the glazing is fitted.

Whilst the focus of this guide is physical security and, in particular, the resistance to manual attack of different generic types of glazing, the guide also contains advice relating to those other performance attributes.

2 Background

The BRE Trust-funded research project 'Security glazing: is it all that it's cracked up to be?' followed a spate of burglaries in the West End of London in 2008. The burglaries in London used 'smash-and-grab' attacks to steal expensive goods such as jewellery and designer clothing from window displays. In one such incident, over half a million pounds worth of jewellery was stolen, with the attack taking place in full view of the public on a busy high street during the middle of the day.

The project involved market research, literature reviews and product testing. That is:

- Crime figures were studied to determine whether glazing was being targeted in a significant portion of burglaries or just unique cases.
- Existing guidance on burglary protection measures and, in particular, security glazing was reviewed to determine the nature of the recommendations it presented in terms of the selection of glazing to protect property from burglars.
- A number of domestic and international standards were examined to establish how security glazing is classified and what tests are conducted in order to validate those classifications.

The main results of the market research and literature reviews are summarised below.

2.1 Glazing in crime

Crime statistics are published annually by the Home Office. Data from the British Crime Survey 2010/11[1], summarised in Appendix A.2, indicated what entry methods were used by burglars.

The data indicated that once figures relating to unsecured doors and windows were omitted, glass was targeted in up to 25% of burglaries on secured premises where the burglar used a degree of force to gain entry (Figure 1). This figure depended on whether the door panels attacked were glazed or were formed from opaque materials such as timber or PVC-U.

Statistics collated within the British Crime Survey 2010/11, together with data available from other sources[3], suggested that attacks on the glazing were second only to those that involved forcing open a complete door or window in order to gain entry. This assumed that statistics involving unsecured doors and windows were excluded from the analysis.

A study by Pascoe and Harrington-Lynn, *The role of windows in domestic burglary*[4], examined the role that windows played in preventing burglary. It found that glass was broken in 29% of burglaries. The glass was broken in one of two ways: either the burglar directly targeted the glass, or the glass broke as a result of loads applied to other elements. Glass within strong and rigid window frames was targeted directly far more often than glass within frames that were formed from less resilient materials.

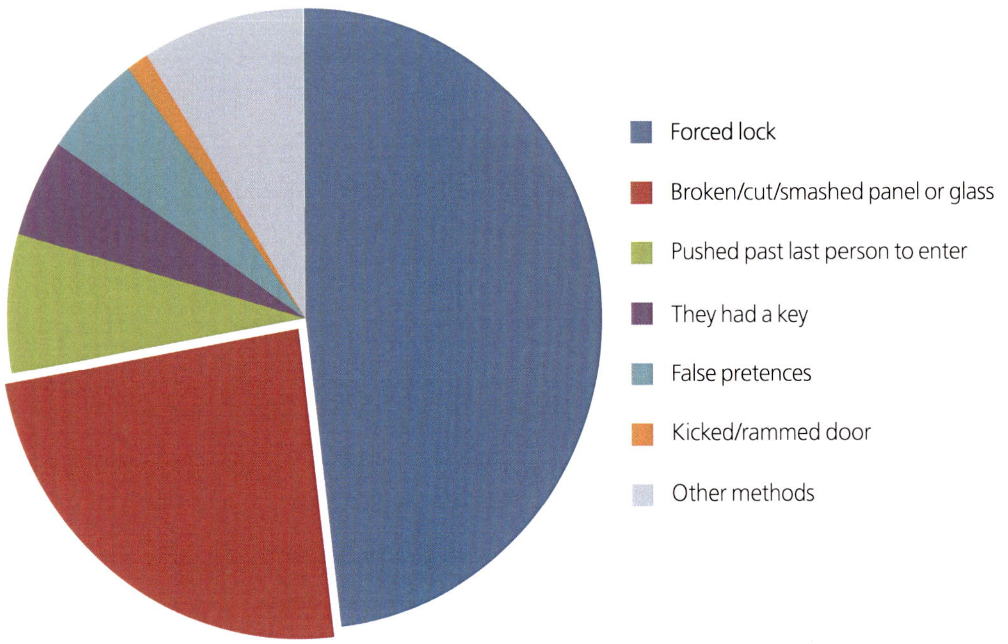

Figure 1: Methods of entry used on windows and doors of domestic properties in England and Wales 2002/03 to 2010/11
Source: *Crime in England and Wales 2010/11: supplementary tables*[2]

It was considered that a number of factors influenced this willingness to attack the glass in the UK. The most important of these was the increased specification of building elements to security performance standards such as BS 7950, incorporating Amendments 1, 2 and 3 (*Specification for enhanced security performance of windows for domestic applications*)[5] and PAS 24 (*Enhanced security performance requirements for door assemblies – Part 1: Single and double leaf, hinged external door assemblies to dwelling*)[6]. These two standards emphasised the need for building elements to resist attempts to lever opening elements open, or remove infill panels, but did not address the vulnerability of the glazing to manual attack.

Implementation of these standards had resulted in frames and locking systems becoming increasingly resilient to manual attacks. However, the glazing's resistance had remained relatively unchanged. This increasingly meant that the glazing was the most vulnerable feature of the windows and doors.

Furthermore, data obtained from several police departments and government organisations in the US suggested that criminals there were increasingly willing to break glass when attempting to enter properties.

The crime figures studied clearly illustrated the need to consider the level of physical security offered by the glazing used within building elements such as windows and doors.

2.2 Guidance on the specification of security glazing

A detailed literature search conducted as part of the research project identified more than 110 UK documents in which measures for tackling domestic burglary crime were discussed. Only 76 of those documents included guidance on physical security measures, and of those, only 32 specifically tackled the subject of glazing.

Examination of these documents revealed that only six recommended the use of a specific type and thickness of glazing. However, it was not made clear on what basis this recommendation was made, or what level of resistance to manual attack this glazing was likely to offer in relation to building elements in which it may be used.

This guide was therefore created to expand upon the existing guidance documents with the addition of relevant performance data in order to validate its recommendations. These performance data were obtained during an extensive test programme conducted on a wide variety of glazing materials by BRE Global Ltd, and are summarised in section 4.2.

2.3 Standards related to the physical security of glazing

Numerous standards associated with glazed building elements are available. Some of these directly tackle the physical security performance of the glazing, whereas many simply state that the glazing should meet the requirements of other standards – for example, BS 7950.

A number of these standards are examined in the appendix. Of all the standards reviewed, only two treated the glazing as an integral component of the building element to which it was fitted. That is, they expected the glazing to offer the same level of resistance to manual attack as the other features of the building element. The two standards identified were LPS 1175 (*Requirements and testing procedures for the LPCB approval and listing of intruder-resistant building components, strongpoints, security enclosures and free-standing barriers*)[7] and SD-STD-01.01 (*Certification standard – Forced entry and ballistic resistance of structural systems*)[8].

Of these, LPS 1175 was supported by LPS 1270 (*Requirements and testing procedures for the LPCB approval and listing of intruder-resistant security glazing units*)[9]. This enabled the glazing's resistance to penetration, and therefore its compatibility with other window/door components, to be determined and specified in its own right.

Glazed building elements tested to other standards cannot be guaranteed to contain glazing that offers the same level of resistance.

A selection of standards related to the manual-attack resistance of glazing are explained in further detail in the following subsections.

2.3.1 LPS 1175

The current version of LPS 1175 (Issue 7) defines eight performance classifications, known as 'security ratings'. These are based upon different levels of threat or risk. Products are only given a security rating if every element of the product (such as hinges of doorsets, glazing and locking hardware) meets the minimum requirements of that security rating.

Security rating 1 (SR 1) requires at least one minute of resistance to attack using basic hand tools that could be concealed in a criminal's pocket, such as small screwdrivers or levers (Figure 2). As the security rating increases, so too do resistance times, the size and complexity of the tools used and the risk involved (Figure 3). For instance, SR 4 requires at least 10 minutes' resistance to small battery-powered tools and larger hand tools such as crowbars, hacksaws and sledgehammers. Further details of the security ratings and tools are available in Table A1 of the appendix.

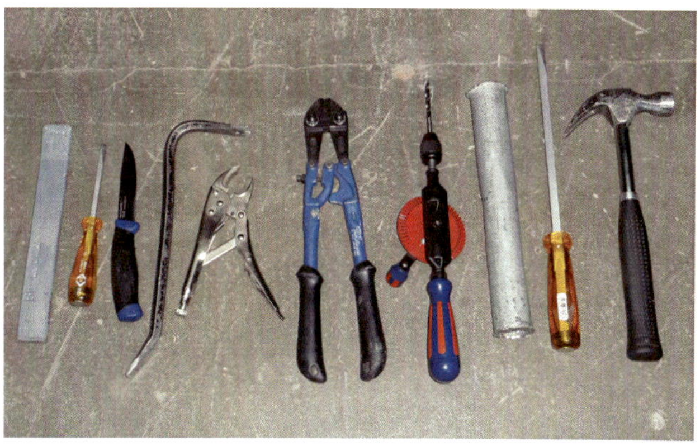

Figure 2: Selection of category A and B tools available to attackers during tests conducted to LPS 1175 security ratings SR 1 and SR 2[7]

Minimum performance required	Notes	Typical applications		
		Domestic	Commercial	Critical national infrastructure
			High risk	
SR 8	Products certified to this rating provide 20 minutes of resistance to extreme means of forced entry into higher value storage areas using a wide range of tools including mains-powered tools.			
SR 7 SR 6	Products certified to these ratings provide 10 minutes of resistance to professional means of forced entry into higher value storage areas using a wide range of tools including mains-powered tools.			
SR 5 SR 4 SR 3	Products certified to these ratings resist experienced attempts at forced entry using a wide range of tools, including battery-powered tools, lasting up to 10 minutes (SR 4 and SR 5) or five minutes (SR 3).			
SR 2	Products certified to this rating provide three minutes of resistance to determined attack by an opportunist burglar using a range of techniques including those that involve creation of noise.			
SR 1	Products certified to this rating provide one minute of resistance to opportunist attempts at entry using a range of techniques including those that involve creation of noise, eg attacks involving breaking glass.			
			Low risk	

Figure 3: Risk relative to security ratings in LPS 1175[7]

2.3.2 LPS 1270

LPS 1270 builds upon the conventions of LPS 1175 and uses the same security ratings and toolkits. However, instead of a single-digit security rating there are three digits, with each digit representing a different failure criterion. For example, a product approved to LPS 1175 may have achieved SR 2, whereas a product approved to LPS 1270 may have achieved SR 224.

The first digit describes the resistance to local penetration. A small hole can be created, through which tools such as a screwdriver or rigid wire may be inserted to operate internal hardware such as a panic bar. The rating may range from 0 to 8, with a 0 indicating that the glazing does not achieve a security rating for this aspect.

The second digit describes the resistance to hand access. A slightly larger hole can be created, through which a criminal may be able to operate devices such as a thumb turn. As with the first digit, the second rating may range from 0 to 8.

The final digit describes the resistance to complete access. This defines the glazing's resistance to the creation of an aperture potentially large enough for a criminal to pass through. The third digit ranges from 1 to 8, and all glazing must achieve at least SR 1 for this aspect in order to achieve a classification to LPS 1270.

2.3.3 BS EN 356

BS EN 356 (*Glass in building – Security glazing – Testing and classification of resistance against manual attack*)[10] is a mechanical test standard, as opposed to a manual test standard like LPS 1175 or SD-STD-01.01. BS EN 356 defines eight performance classifications.

The lowest five classifications, P1A to P5A, are determined by dropping a specified steel sphere on three points of an equilateral triangle in the centre of the glass.

The higher classifications, P6B to P8B, subject the glass to a series of mechanically operated hammer and axe strikes. The glazing must resist the creation of a 400 mm square hole within the number of strikes defined by the classification.

Further details of the classifications are available in Table A2 of the appendix.

3 Types of glazing

Glazing for windows and doors comes in a huge variety of forms. Some glazing contains strands of wire embedded between two sheets of glass to electrically heat it, activate an alarm system when the glass (and wire within) is broken, or to hold the glass together when broken. Some types contain special coatings to make them more, or less, reflective, whereas other coatings can react to ultraviolet light from the sun in order to break down dirt on the surface of the glass.

The following section provides an overview of the typical glazing that may be specified for use in glazed building elements, for instance by many of the other existing guidance publications. They may have optional coatings or special treatments such as the ones listed above, but their primary focus would be physical security.

3.1 Acrylic

Acrylic, like polycarbonate (see section 3.7), is a transparent plastic. Although it is stronger than glass, its strength does not quite equal that of polycarbonate. A sheet of acrylic will weigh roughly half that of a sheet of glass of equivalent dimensions.

Due to its weight and strength advantages over glass, it has also been used as a glass substitute in recent years, eg in motorcycle helmet visors or lenses for exterior vehicle lights. However, acrylic is easily scratched and will typically cost more than an equivalent sheet of glass.

Pieces of acrylic can be chemically welded together. The seams 'melt' into a single piece of solid material, and once polished are almost invisible. Acrylic may be cut to size using conventional wood- or metalworking tools, whereas float glass requires specialist tools to score it prior to snapping it.

3.2 Fire-resistant glass

Fire-resistant glass provides a barrier against the effects of a fire for a defined period of time. It is available in two types: 'integrity only' and 'integrity and insulation'.

'Integrity only' fire glass acts to contain flames, smoke and fire gases, but does not prevent the heat created by the fire from being radiated through the glass. Wired glass (see section 3.9) is a prime example, with the wire embedded within the glass helping it to retain its structure. Other 'integrity only' glasses include ceramic, toughened soda lime silicate and toughened borosilicate.

'Insulating' fire glass builds upon the containment offered by 'integrity only' glass, but also prevents heat transfer by radiation. Special laminates, incorporating gel or intumescent interlayers, offer insulating properties. The layers within these laminates react to the heat of the fire and act as an insulator as they become opaque.

Fire-resistant glass is classified based upon the time in minutes until failure occurred in a standard fire-resistance test furnace. The classification will indicate whether it was tested for integrity (E) and/or insulation (I). For instance, 'E15' glass achieved an integrity performance of 15 minutes, whereas 'EI60' glass achieved both an integrity and insulation performance of 60 minutes.

3.3 Float glass

Float glass, also referred to as 'annealed glass', is probably the most common type of glazing used in buildings the world over. The majority of float glass is produced by the process invented in the 1950s, in which molten glass is poured onto a bath of molten metal – usually tin. The molten glass then settles and floats on top of the tin. As it levels and spreads out along the bath, a smooth face is formed on both sides of the ribbon of glass.

As the ribbon of glass cools, considerable stresses develop that make it liable to crack or shatter when subjected to relatively small sudden temperature changes (thermal shocks) or mechanical shocks. In order to relieve these stresses and reduce the risk of failure of the glass, it is then put through an annealing process whereby it is slowly cooled. This cooling can be several tens of degrees per hour for thin sections of glass, or fractions of a degree per hour for thick slabs or castings of glass.

Annealed glass has a tendency to break into large, jagged shards (Figure 4), which can cause serious injury. Because of this, the use of annealed glass in many parts of the world is restricted, so it cannot be used in areas where there is a high risk of it breaking and causing injury (for instance, at low heights and in bathrooms, fire exits and schools).

Figure 4: Breaking pattern of float glass

In terms of the UK, Approved Document N: Glazing[11] details the impact performance required from glass depending on its intended location in a building.

The European standard for float glass is BS EN 572-2 (*Glass in building – Basic soda lime silicate glass products – Part 2: Float glass*)[12].

3.4 Glass brick

Glass bricks, also known as 'glass blocks', were developed in the early 1900s to provide natural light in industrial buildings. They can be used to form solid structures and physical barriers just as with regular bricks, but with the addition of allowing for privacy or visual obscuration (due to patterns within the glass) while admitting light through.

Modern forms of glass brick tend to be constructed in two halves and then joined together, to form a hollow unit with a slight vacuum. This vacuum offers improved acoustic and thermal insulation over the solid type of glass brick.

3.5 Insulated glazing units

Insulated glazing units comprise multiple panes of glass (also referred to as 'lites') separated by a small gap and assembled into a sealed unit (Figure 5). 'Double glazing' is a prime example of an insulated glazing unit.

The panes are separated by a metallic or composite spacer bar and the cavity between the panes may be filled with air or gas (eg argon) in order to improve the insulation of the unit. A primary seal (normally butyl) may be applied to the edge of the spacer before application to the panes. Once assembled, a secondary seal is applied around the perimeter of the assembled unit. This secondary seal can be polysulphide, polyurethane, silicone or butyl.

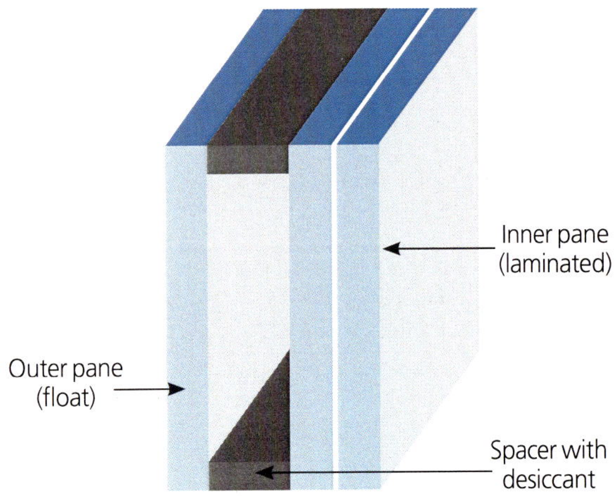

Figure 5: Cross section of an insulated glazing unit

The spacer bar is hollow and filled on two edges with microbead desiccant, which dries the air in the now-sealed cavity via a row of small holes in the spacer bar. This prevents the formation of condensation on either of the glass surfaces facing into the cavity.

The European standard for insulated glazing units is BS EN 1279-1 (*Glass in building – Insulating glass units – Part 1: Generalities, dimensional tolerances and rules for the system description*)[13].

3.6 Laminated glass

Laminated glass consists of two (or more) layers of glass bonded together with an interlayer. This interlayer is either a liquid resin or a foil (sheet) of materials such as polyvinyl butyral (PVB) or ethylene-vinyl acetate (EVA). The layers are sandwiched together to create a monolithic sheet of glass.

With PVB or EVA interlayers, the layers are assembled by heating and pressuring the sandwich in either an autoclave or vacuum bag and oven, creating a predominantly physical bond between the glass surface and plastic interlayer. With a liquid-resin interlayer, the liquid is metered and injected into a cavity created between the sheets, where it then catalytically cures to create a physical and chemical bond between the glass and the resin.

Any type of glass may be bonded together to form laminated glazing, from float glass to toughened glass to polycarbonate. The combination of different materials results in advantages over the individual types of glazing.

When one layer of glass is broken, the shards will typically remain bonded to the interlayer while the remaining layers provide stability and rigidity, keeping the glazing in place within its frame.

As well as providing varying degrees of safety (depending on the type and thickness of the glass and interlayer), the interlayer can also enhance other physical attributes such as acoustic attenuation. The effectiveness of the bond and other properties (eg acoustics) is determined by the type of interlayer used.

The European standard for laminated glass is BS EN ISO 12543-1 (*Glass in building – Laminated glass and laminated safety glass*)[14].

3.7 Polycarbonate

Polycarbonate is a transparent plastic resin and is one of the toughest transparent materials. It is far more commonplace than one might expect due to its use in the optical media market to produce CDs and DVDs.

Compared with a sheet of glass with similar dimensions, a sheet of polycarbonate is stronger and far more difficult to break by impacting. It is also approximately half the weight of glass of the same dimensions. It is often touted as being 'unbreakable' because of its strength and ability to undergo significant deformation before it cracks or breaks. However, this ability depends on its thickness, the quality of the raw material and the manner in which it is processed.

Because of its strength, weight and optical properties, polycarbonate has also been used to create the so-called 'bubble helmets' used by the Apollo astronauts, and more recently, the canopy of the F-22 fighter jet.

Although tough, polycarbonate is not a 'hard' material and will easily scratch. Regular interaction, such as people accidentally brushing the polycarbonate with their keys as they walk past or frequent cleaning with abrasive materials, can result in the build-up of scratches, which reduce the clarity of the glazing.

The risk of scratching can be reduced if the polycarbonate is covered with a film or coating (known as a 'hard coat'), laminated with another material or used in a location that is not prone to such regular interactions.

Older manufacturing processes tended to result in a yellowing over time, as the polycarbonate reacted to ultraviolet light and degraded. Applying a 'hard coat' and adopting newer manufacturing processes can reduce this effect, resulting in a much longer lifespan for polycarbonate glazing. However, contamination through scratches can still lead to embrittlement and cracking of the polycarbonate under impact.

Polycarbonate melts at a much lower temperature than many other materials used in glazing. If un-coated or not used as part of a laminate, its ability to burn means it could be vulnerable to acts of vandalism. Furthermore, polycarbonate is highly susceptible to solvents and may be cut using a variety of hand tools.

3.8 Toughened glass

Toughened glass (also referred to as 'tempered glass') is manufactured from float glass that has been strengthened by controlled thermal or chemical treatments.

Float glass is reheated then rapidly cooled. As a result, the surfaces of the glass shrink and initially develop tensile stresses. When the core of the glass begins to cool, it contracts, which forces the already-solidified surfaces to contract. This results in compressive stresses on the surface while the core (approximately 60% of the thickness of the pane) develops tensile stresses (Figure 6). The compressed surface stresses contribute to the greater strength of toughened glass over float glass.

The term 'toughened glass' is generally used to describe fully tempered glass, but is also sometimes used to describe heat-strengthened glass. Heat-strengthened glass is twice as strong as annealed glass, whereas fully tempered glass will typically be five times as strong.

Unlike other types of glass, if any part of tempered glass fails the entire panel will shatter at once into small fragments (Figure 7) that are less likely to cause injury than the shards produced when float glass is broken.

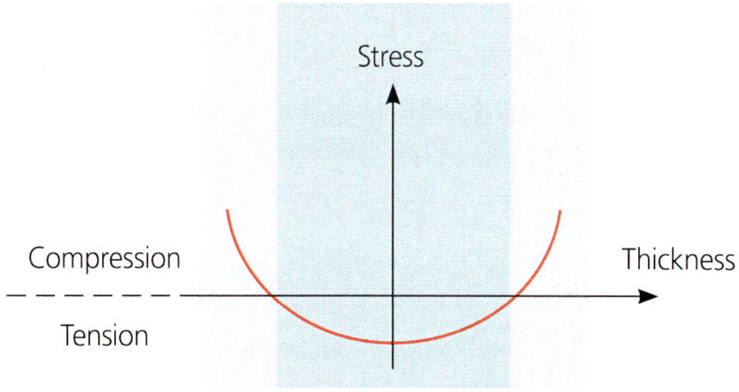

Figure 6: Distribution of stresses within a pane of toughened glass

Figure 7: Breaking pattern of toughened glass

Toughened glass is most susceptible to shattering if damage occurs at its edges or if the impact is concentrated, such as the surface of the glass being struck with a centre punch. Toughened glass will also deform more than annealed glass when subjected to the same loads.

While annealed glass can be scored after undergoing the annealing process, toughened glass cannot be re-worked after going through the toughening process. It must be cut to size or pressed to shape, the edges polished and any holes need to be drilled before toughening.

The European standard for thermally toughened soda lime silicate glass is BS EN 12150-1 (*Glass in building – Thermally toughened soda lime silicate safety glass*)[15].

3.9 Wired glass

Wired glass (eg Georgian wired polish plate or Georgian wired cast glass) is manufactured by embedding a steel wire mesh (typically of 10–12 mm spacing; see Figure 8) between two semi-molten ribbons of glass. These ribbons of glass are then pressed together to form a single sheet.

A common misperception of wired glass is that the wires increase its resistance to impacts. The presence of the wires actually weakens the glass, which works to reduce its impact resistance. However, when the glass is broken the wire mesh helps to retain the shards for safety.

Wired glass is commonly used as a basic fire-resistant glass, and was for many years the only type of glass that could pass the requirements for fire-resistance testing. New forms of glass have since been developed that can outperform wired glass and achieve much longer fire-resistance ratings.

Figure 8: Wire mesh embededed in a pane of float glass

4 The selection of effective security glazing

4.1 A step-by-step process

Glazing comes in a wide range of different varieties, many of which were described in section 3. But how can specifiers ensure that the glazing they have selected for a secure building element is the right choice?

By following an appropriate sequence of selection (Figure 9), specifiers can ensure that the glazing will offer an appropriate level of resistance to attack by criminals. The sequence also takes into account aspects of the glazing such as acoustic insulation, thermal insulation or fire-resistance performance. Although not necessarily related to the physical security of the building element, these additional performance attributes may be desirable.

This guide focuses primarily on resistance to physical attack. As such, emphasis has been placed on determining the security requirements above additional properties such as acoustic insulation. If desired, it is possible to substitute the primary requirement with alternative requirements to ensure the effective selection of glazing for other uses.

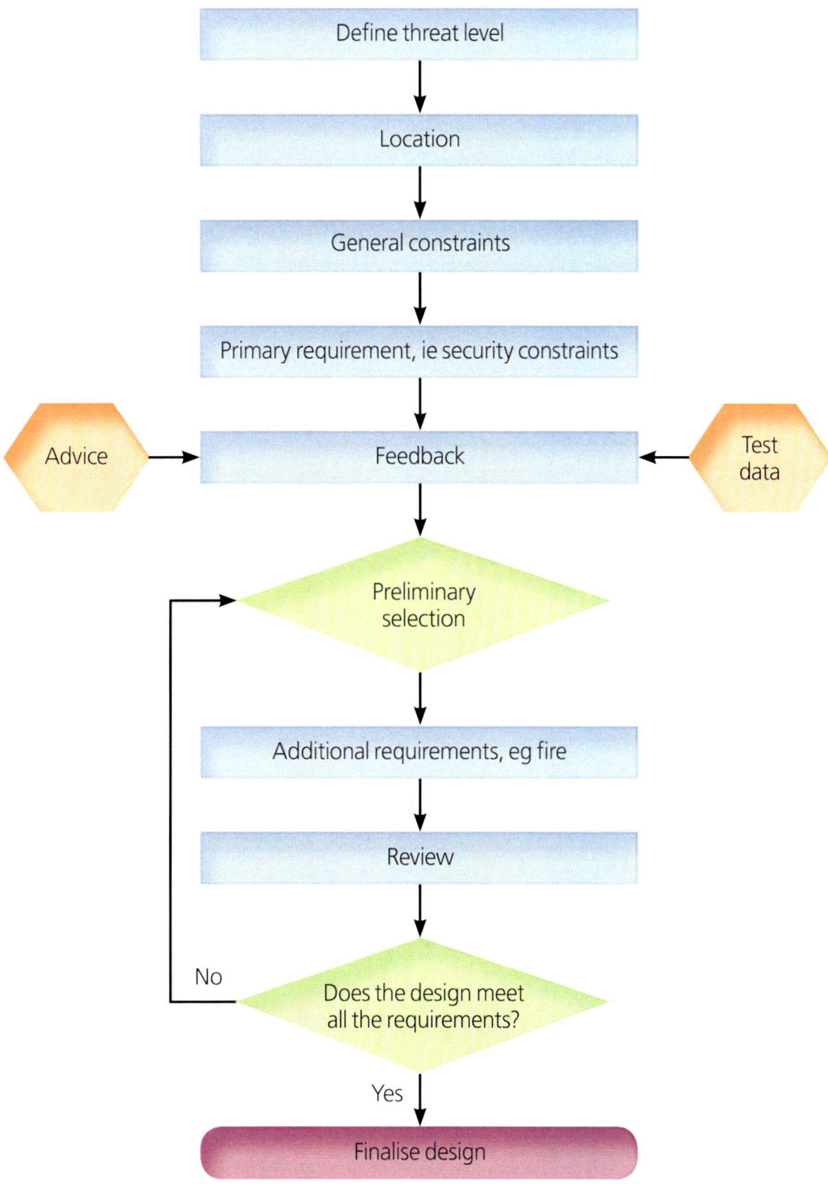

Figure 9: Sequence of selection of effective glazing for use in secure building elements

4.1.1 Location and general constraints

When selecting glazing of any sort, the most logical place to begin is its location. What kind of building is it? Is it a new building under construction, or an existing building being refurbished and renovated? Where in the building will the glazing be used, and which building element (eg doorset, window) will it be fitted to?

The answers to these questions will feed into a number of constraints. For instance, there may be size or weight limits imposed by the location. Fitting a heavier type of glazing into a lightweight aluminium doorset after a break-in, for example, may cause the hinges used to support the door to suffer excess wear or even break due to the increased weight.

Invariably, there will be budgetary constraints, which will need to be considered as part of an overall cost-benefit risk-based process. There may also be an associated time constraint, such as completion in time for a major event.

The properties of a number of generic types of glazing are compared in Table 1. Despite each sample having similar dimensions, the mass and relative costs differ considerably. However, mass and cost should not be the only properties of glazing considered. Although more than twice the price of float glass, polycarbonate will provide significantly more resistance to manual attack than float glass of an equivalent thickness. Refer to section 4.2 for more information on manual attack resistance.

4.1.2 Security

The security employed around a building's facades will vary depending on the nature and contents of the building. For instance, government buildings or banks may require a level of resistance for physical attack, explosions and ballistic attacks, whereas a local retailer may only require resistance to burglary and vandalism. Furthermore, different degrees of protection exist. For instance, a jewellery store with high-value goods will require a greater level of security than a local newsagent.

Specifiers should define the threat level, and what degree of security is therefore appropriate for the building. Although the emphasis may be on resistance to physical attack, alternative security requirements such as explosion or ballistic protection may be considered at this stage. Refer to sections 4.3.1 and 4.3.2 for further information.

Standards such as LPS 1175 (described in section 2.3.1) allow specifiers to select the performance according to the risk. For example, the police or insurers may recommend that the facade of a jewellery store offers resistance commensurate with the requirements of SR 4. Meanwhile, they may consider the requirements of SR 2 to be sufficient for a local newsagent.

Once an appropriate level of security has been decided upon, attention should then be turned to the locking hardware of doorsets and windows. This can pose serious security concerns, particularly when glazing is fitted. The external faces of these building elements typically require a key in order to be unlocked and opened, or may offer no means of entry into the building at all.

However, the internal face may be fitted with a variety of hardware operated by devices such as keys, thumb turns or emergency and panic exit hardware. A thumb turn on the internal face does not require a key to unlock its hardware; turning the device unlocks it and allows it to be opened when a handle is operated. Emergency exit hardware such as lever handles, push pads and panic bars permit users to unlock and open building elements such as doors in a single action to allow for a timely exit in case of fire or other emergencies. This means that a criminal could make a relatively small hole in the glazing, through which they could pass tools such as a piece of wire or a hook, rotate it and pull it back to operate the panic bar. This would open the door and allow the criminal complete access to the contents of the room.

Table 1: Comparison of properties of different generic types of glazing

Glazing type	Width (mm)	Height (mm)	Thickness (mm)	Mass (kg/m^2)	Relative cost*
Float glass	865	1100	10	25	£
Toughened glass	865	1100	10	25	££
Polycarbonate	865	1100	10	12.2	£££
Laminate (toughened and polycarbonate)	865	1100	10	17	££££
Laminate (float and PVB)	865	1100	11.5	26.5	££
Laminate (toughened and PVB)	865	1100	11.5	26.5	£££
Laminate (toughened and PUR)	865	1100	12	27.1	£££
Laminate (fire)	865	1100	11.4	25	££££

* Relative cost is represented on a scale from £ (least expensive) to £££££ (most expensive).
PVB Polyvinyl butyral.
PUR Polyurethane.

If, however, the door was not an emergency exit, key hardware could be fitted to the inside and a criminal would therefore have to create a much larger hole in the glazing to allow them to pass through completely, as they would not have the required key to unlock the door.

Standards such as LPS 1270 and LPS 1175 consider a criminal's ability to exploit smaller holes. LPS 1175 and SD-STD-01.01 were also the only two standards identified during the research that required the glazing to offer the same resistance to attack as the rest of the building element in which they were fitted.

If glazing cannot be found that offers the correct level of protection for the locking hardware used on the building element, the choices are limited. Either the glazing can be removed or restricted (although care should be taken to ensure compliance with relevant building regulations), or the choices made during the first two stages of the sequence defined in Figure 9 can be re-examined. If glazing is still required, then its size, shape or location should be reconsidered.

For example, if a doorset incorporated a thumb turn on the internal face but it was not possible to source glazing with adequate resistance to hand access, then consideration should be given to relocating the glazing further from the thumb turn. This way, although criminals may be able to reach through a small hole, the hole should then be too far away from the thumb turn to allow them to operate it. Alternatively, the thumb turn could be shrouded to prevent direct access to it after reaching through such a hole in the doorset.

Another example is if a doorset incorporating key locking hardware on both faces could not be fitted with glazing that offered sufficient resistance to complete access. In this case, the glazed area should be made smaller such that even if it were entirely removed by the criminal the resulting aperture would be too small for them to pass through. The elliptical test block shown below represents the area needed to allow complete access through the doorset (Figure 10).

However, it is important to recognise that it is not just the type and location of glazing that determines the level of protection offered, but also its installation. The size of the rebate and

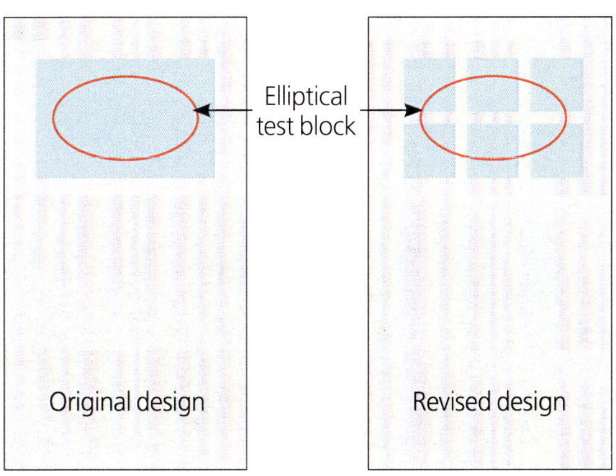

Figure 10: Redesigning glazing with building elements in order to restrict access

glazing beads, materials and fixings all contribute to the security of glazing units. Furthermore, beads should be located on the defended side wherever possible. This is because those positioned on the 'attack side' are far more prone to removal.

4.1.3 Feedback and preliminary selection

Once all the constraints have been taken into account, it may be prudent to consider feedback from previous projects that may have been similar. Lessons learned during those projects may be implemented to aid the current project.

It may also be wise at this stage, or earlier in the sequence if desired, to seek advice from independent organisations such as Secured by Design (SBD), the Loss Prevention Certification Board (LPCB) or the Association of British Insurers (ABI).

4.1.4 Additional requirements

Once an initial selection of glazing has been made or a number of suitable types of glazing identified, the additional properties they offer can be examined. Although the emphasis has thus far been on physical security, glazing has come a long way since the form seen in early Mesopotamia; it is now much more than just a means to keep out the elements while admitting daylight.

The project in question may have additional requirements, such as offering a level of fire resistance to provide employees or customers with a safe route or means of escape during a fire.

Further guidance on these other performance attributes is provided in section 4.3.

Other requirements to consider at this stage are maintenance issues, design requirements for natural lighting and weather-resistance issues such as wind loading and watertightness.

4.1.5 Review and finalise design

Fulfilling all the requirements is a delicate balance between the correct level of protection, cost, the aesthetics of the building and compliance with building regulations: an architect may prefer fully-glazed curtain walls whereas a blast engineer may prefer no glazing at all; there will be regulatory requirements to be met; and, of course, the client will have cost requirements.

Unfortunately, finding materials that meet all the requirements for the 'right price' is not always possible. Some may meet only one requirement while others may achieve a number of them.

In the example illustrated overleaf, product 'A' meets all the requirements indicated and would therefore be the obvious choice (Figure 11). However, if there was a supply problem that prevented product 'A' from being sourced in time for a project, it would be necessary to select an alternative product that was the 'next best fit'. In the example shown, product 'B' achieves all but the explosion resistance requirements. If that product was to be specified in place of product 'A', it would be necessary to take into account the need to implement additional mitigation measures for that risk. Such measures may make the use of product 'B' impractical or uneconomic, forcing another alternative to be selected.

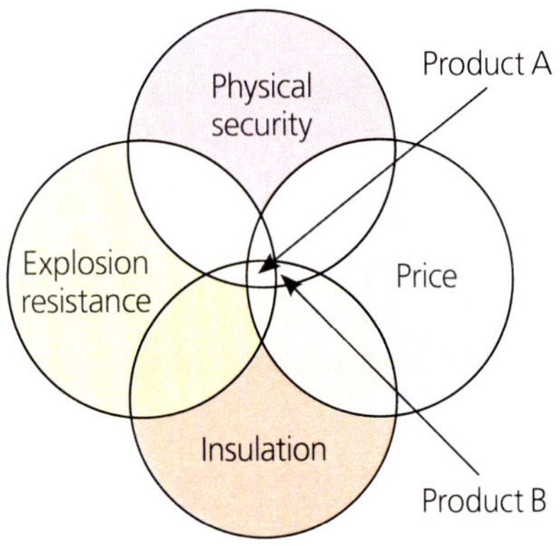

Figure 11: Example product performance factors

4.2 What performance data are available?

Sections 2.2 and 2.3 discussed numerous pieces of guidance and standards that attempt to tackle glazing in secure building products. Many of these simplify the issue of glazing performance by containing statements such as 'The glazing must conform to the requirements of BS 7950'. However, it is not clear what performance is actually expected of the glazing.

As part of a programme of research funded by BRE Trust and conducted by BRE Global Ltd, a number of generic specimens of glazing were obtained for the purposes of testing and determining their resistance to manual attack. The manual attack tests conducted were based on the use of the tools and working times defined in the following standards: BS 7950, PAS 24, LPS 1175, LPS 1270 and BS EN 1627 (*Pedestrian doorsets, windows, curtain walling, grilles and shutters – Burglar resistance – Requirements and classification*)[16]. That is, the tools used to manually attack other aspects of the building elements (such as frames and mid-rails) were used to attack the glazing. Although the tests defined in BS 7950, PAS 24 and BS EN 1627 would not normally be conducted in this way, the tests enabled BRE Global Ltd to directly compare the performance of the glazing specimens with each other and with the performance expected of the remainder of the building element within which the glass may be used.

As discussed in section 4.1.2, different sizes of aperture can be exploited to operate different types of locking hardware on the internal face of windows and doors. Because of this, the time taken to create an aperture through the glazing for different size test blocks was recorded.

The test blocks detailed below (Table 2) were defined in the standards as being representative of a typical criminal and their tools being able to exploit different aspects of the building element.

The following sections summarise the results for different types of access aperture to illustrate the variation in resistance afforded by a selection of different generic types of glazing (Table 3). The configurations tested were a selection of those available on the market; numerous other configurations were available, which were not tested (eg float glass and polycarbonate laminates).

4.2.1 Complete access holes

The test results are shown overleaf for different thicknesses of laminated glazing formed from layers of float glass and PVB (Figure 12). Some important conclusions can be drawn from Figure 12.

Firstly, tests on the same specimen of glass to different standards can give an alarming spread of results. This is especially evident from the thicker glazing units, where the resistance times achieved during tests to LPS 1175 and LPS 1270 were approximately half those achieved during tests to BS EN 1627.

Table 2: Test blocks

Access	Shape	Major axis (mm)	Minor axis (mm)	Applicable to which standards?
Local	Circular	25	25	BS 7950 and PAS 24
	Rectangular	25	8	LPS 1270
Hand	Circular	50	50	BS 7950* and PAS 24
	Circular	60	60	LPS 1270
Complete	Elliptical	380	225	BS 7950 and PAS 24
	Circular	350	350	BS EN 1627†
	Elliptical	400	225	LPS 1175 and LPS 1270

* BS 7950 does not specify this test block but it was used during the tests conducted by BRE Global Ltd.
† BS EN 1627 also specifies elliptical (400 mm × 300 mm) and rectangular (400 mm × 250 mm) test blocks but these were not used during the tests conducted by BRE Global Ltd.

Table 3: Test specimens

Glazing type	Thickness (mm)	Construction (attack face listed first)
Wired glass	6	6 mm float glass with encapsulated mild steel wires
Float glass	10	10 mm float glass
Toughened glass	4	4 mm toughened glass
Laminate (float/PVB)	6.4	3 mm float glass, 0.38 mm PVB interlayer, 3 mm float glass
Laminate (float/PVB)	11.4	2.8 mm float glass, 1.5 mm PVB interlayer, 2.8 mm float glass, 1.5 mm PVB, 2.8 mm float glass
Laminate (float/PVB)	11.5	3 mm float glass, 1.5 mm PVB interlayer, 3 mm float glass, 1.5 mm PVB, 3 mm float glass
Laminate (toughened/ polycarbonate)	16	4 mm toughened glass, 2 mm PUR interlayer, 4 mm polycarbonate, 2 mm PUR, 4 mm toughened glass
Laminate (toughened/ polycarbonate)	18	4 mm toughened glass, 2 mm PUR interlayer, 6 mm polycarbonate, 2 mm PUR, 4 mm toughened glass
Laminate (fire)	11.4	3 mm float glass, 1.2 mm intumescing resin interlayer, 3 mm float glass, 1.2 mm intumescing resin, 3 mm float glass

PVB Polyvinyl butyral.
PUR Polyurethane.

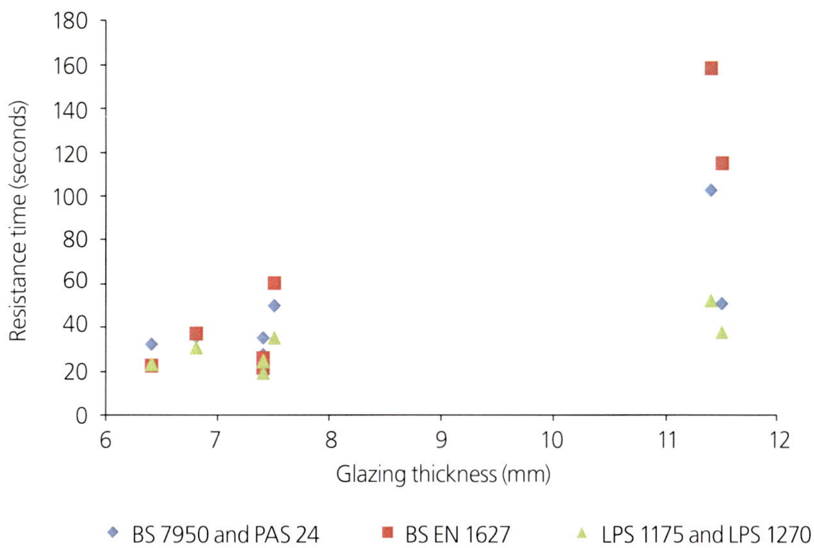

Figure 12: Attack resistance times for different thicknesses of laminated glass (float/PVB)

This is a reflection of the tools specified within those standards and their effectiveness at penetrating the glass. It highlights the necessity to identify the types of tools the glazing needs to resist, and the need to ensure that the standard and grade specified include those tools.

Secondly, the glazing's resistance, and therefore the protection provided, do not correlate with its thickness. Thus, specifying glass by thickness alone is not an effective method of ensuring that the glazing provides the required level of protection.

Thirdly, none of the specimens achieved the resistance times required by the remainder of the building elements for each of the standards. For instance, a doorset tested to PAS 24 is required to withstand three minutes of manual attack, yet none of the laminated glazing detailed in Figure 12 achieved anywhere near three minutes of resistance. This was despite PAS 24 and many published pieces of guidance suggesting that 6.4 mm laminated glass would be suitable.

Furthermore, products are required to offer at least one minute of resistance for a complete access aperture in LPS 1175 and LPS 1270. Only one of the specimens approached that level of resistance.

The test results on the specimens described in Table 3 are summarised in Table 4. Glazing formed from polycarbonate (either monolithic polycarbonate or as part of a glass/polycarbonate laminate) was the only type to offer resistance times consistent with those of the remaining elements that make up doors and windows classified in accordance with those standards.

As mentioned in section 2.2, most existing guidance documents specified the use of laminated glazing. However, they did not specify its composition.

It is clear from the results shown in Table 4 that the glazing recommendations from existing documents would not offer the same level of resistance to attack as the rest of the building element unless it incorporated polycarbonate or was significantly thicker than 11.5 mm. Furthermore, the thickness and number of interlayers can have a positive effect on the attack resistance.

4.2.2 Hand-sized access holes

The results summarised in Table 5 are of attempts to create hand-sized access holes through the same range of glazing materials as those discussed in the previous section (ie Table 3). Such a hole would potentially enable internal hardware such as thumb turns and lever handles to be operated.

There is a dramatic reduction in the resistance times compared with those of Table 4 because a significantly smaller aperture is formed through the glazing.

A noteworthy result applies for those who specify doorsets and windows to meet the requirements of PAS 24. This standard dictated that doorsets fitted with non-key-operated hardware such as thumb turns (which could be exploited through the holes created above) should be glazed with laminated glass.

However, the laminated glass tested (6.4 mm thick) could be overcome in as little as five seconds. Even much thicker PVB-based laminates, such as the 11.4 mm thick variety, could be overcome in less than 16 seconds using the tools defined in PAS 24. This is less than one-tenth of the time that the remainder of the door is required to resist if attacked using those tools, in order to achieve a classification to PAS 24!

Table 4: Time to create aperture through different types of glazing – complete access

Glazing type	Thickness (mm)	Attack resistance time (seconds)*		
		BS 7950 and PAS 24	BS EN 1627	LPS 1175 and LPS 1270
Wired glass	6	69	28	19
Float glass	10	21	~ 2	~ 2
Toughened glass	4	~ 2	~ 2	~ 2
Laminate (float/PVB)	6.4	33	23	33
Laminate (float/PVB)	11.4	103	159	52
Laminate (float/PVB)	11.5	51	115	38
Laminate (toughened/polycarbonate)	16	> 180	> 180	203
Laminate (toughened/polycarbonate)	18	> 180	> 180	314
Laminate (fire)	11.4	151	29	17

* Times in red are below the resistance time required to achieve a classification in accordance with the relevant standards.
PVB Polyvinyl butyral.

Table 5: Time to create aperture through different types of glazing – hand-sized access

Glazing type	Thickness (mm)	Attack resistance time (seconds)*	
		BS 7950 and PAS 24	LPS 1175 and LPS 1270
Wired glass	6	11	4
Float glass	10	< 21	~ 2
Toughened glass	4	~ 2	~ 2
Laminate (float/PVB)	6.4	5	4
Laminate (float/PVB)	11.4	16	10
Laminate (float/PVB)	11.5	8	7
Laminate (toughened/polycarbonate)	16	> 180	38
Laminate (toughened/polycarbonate)	18	> 180	271
Laminate (fire)	11.4	24	3

* Times in red are below the resistance time required to achieve a classification in accordance with the relevant standards.
PVB Polyvinyl butyral.

4.2.3 Local access holes

The results listed in Table 6 overleaf are of attacks aimed at creating the smallest size aperture through the glass. It may be possible to exploit such holes to pass through tools such as pieces of wire, hooks or small levers and screwdrivers to operate devices such as panic bars or lever handles.

Just as before, laminated glazing incorporating polycarbonate offered the most resistance to attack. However, its resistance was greatly affected by the thickness of the polycarbonate used.

4.2.4 Summary of performance

The performance of different generic types of glazing during tests conducted by BRE Global Ltd is summarised in Table 7 overleaf. It describes whether different generic types of glazing are potentially suitable for use in windows and doors with different internal locking hardware.

The information is provided only as an illustrative guide, and any performance declared by manufacturers should be confirmed by independent verification wherever possible.

4.3 Additional features

As previously mentioned, glazing can do much more than just keep the weather (and criminals) out. Some of its additional characteristics are discussed below. For some aspects, such as fire resistance and explosive protection, far more in-depth guidance is available, as listed in section 6.

4.3.1 Ballistic protection

Specialist glazing can also be used to reduce the risk of injury during ballistic attacks. This is often seen in banks with tellers sitting behind thick sheets of glass, or used in armoured vehicles to protect VIPs.

Stopping a bullet is similar to catching a fast-moving ball, although the bullet travels at much greater velocities. When catching a ball, the trick is to move your hand back and slow the ball down gradually, changing its momentum slowly enough to reduce the force you feel. The same principle applies to ballistic protection, changing the momentum of the bullet.

Ordinary glazing cannot bend or absorb a bullet's kinetic energy, so will shatter and allow the bullet to continue with negligible change in momentum. Furthermore, the shards formed by the impact will act as additional projectiles and may therefore cause injuries to anyone in the vicinity of the rear of the glazing.

Bullet-resistant glazing can reduce the bullet's kinetic energy due to the combined action of the laminates used to form it (up to 120 mm thick in some cases). As the bullet strikes the glazing, its energy is spread out across each layer. Being divided among layers and spread over a large area, the energy is quickly absorbed such that the bullet has insufficient energy to continue piercing the layers.

The interlayers will usually work to retain fragments when the glass is broken. The thickness and number of layers generally determine how much energy the glazing can absorb. However, when the protected face is not a material such as polycarbonate, there is still a risk of spalling. This could result in fragments of glass being ejected and causing injuries.

Table 6: Time to create aperture through different types of glazing – local access

Glazing type	Thickness (mm)	Attack resistance time (seconds)*	
		BS 7950 and PAS 24	LPS 1175 and LPS 1270
Wired glass	6	6	1
Float glass	10	< 21	~ 2
Toughened glass	4	~ 2	~ 2
Laminate (float/PVB)	6.4	3	2
Laminate (float/PVB)	11.4	8	3
Laminate (float/PVB)	11.5	4	2
Laminate (toughened/polycarbonate)	16	> 180	13
Laminate (toughened/polycarbonate)	18	> 180	95
Laminate (fire)	11.4	12	1

* Times in red are below the resistance time required to achieve a classification in accordance with the relevant standards.
PVB Polyvinyl butyral.

Bullet-resistant glazing is rated for different calibres of projectile and weapons. For instance, bullets fired by a rifle have generally greater kinetic energy than those fired by a handgun. Therefore, thicker glazing is required if rifle attacks are expected.

Ratings for ballistic protection are detailed in BS EN 1063 (*Glass in building – Security glazing – Testing and classification of resistance against bullet attack*)[17]. This is summarised in Table A3 of the appendix. Apart from the SG1 rating, the glazing is subjected to multiple shots. Although the first bullet may be stopped, the damage caused by the bullet may weaken the glazing surrounding the first shot and potentially let subsequent shots through.

Recent advancements include:

- the production of cast polyurethane resin, which offers similar properties to polycarbonate
- the invention of one-way bullet-resistant glazing, used on some armoured vehicles.

With one-way bullet-resistant glazing, bullets fired at the vehicle will be stopped as with regular bullet-resistant glazing, but occupants of the vehicle will be able to shoot out at their assailants. This is achieved through the lamination of a brittle outside layer and a flexible internal layer.

Bullets fired at the brittle layer shatter it, absorbing some of the bullet's energy and slowing it. The bullet then encounters the flexible layer and is stopped. However, bullets fired from the inside encounter the flexible layer first and penetrate (because the energy is focused on a much smaller area) through to the brittle layer without significantly hindering the bullet's progress.

4.3.2 Explosion protection

Modern buildings make far greater use of glazing than older ones. This, combined with the properties of the glazing itself, means that the greatest cause of injuries from an explosion nowadays is usually the fragmentation and projection of glass.

Measures designed to reduce these risks have been put in place by the UK government as part of its fight against the threat of terrorism.

When a bomb explodes, it creates an almost instantaneous rise in pressure. As this positive pressure decays, the air forced out by the explosion tries to rush back in and fill the void left behind, creating a negative pressure. This process is shown in Figure 13.

The peak and duration of the positive pressure phase result in the positive specific impulse (the shaded area of the graph). These values feed into the performance classifications of glazing to BS EN 13541 (*Glass in building – Security glazing – Testing and classification of resistance against explosion pressure*)[18], which are summarised in Table A4 of the appendix. Additionally, the classifications may be supplemented with an 'S' (Spalling) or 'NS' (Non-spalling) suffix to denote whether splinters originate from the protected face of the glazing when tested.

The sudden changes in pressure can cause glazing to fail catastrophically. Float glass will break into dagger-like shards, which can be thrown as extremely hazardous projectiles into the building and surrounding areas during an explosion. Toughened glass breaks into safer 'dice' but these may travel at higher velocities than shards of float glass, still resulting in injuries deep within a building. Other injuries such as perforated eardrums arise from the shock waves of the explosion itself.

Table 7: Use of glazing with different internal locking hardware

Glazing type	Thickness (mm)	Internal locking hardware		
		Key	Thumb turn/ lever handle	Emergency/panic exit
Wired glass	6	×	×	×
Float glass	10	×	×	×
Toughened glass	4	×	×	×
	6	×	×	×
	10	×	×	×
Polycarbonate	6	✓✓✓	✓✓	✓
	8	✓✓✓	✓✓	✓✓
	10	✓✓✓	✓✓	✓✓
Laminate (float/PVB)	6.4	×	×	×
	6.8	×	×	×
	7.4	×	×	×
	7.5	×	×	×
	11.4	×	×	×
	11.5	×	×	×
Laminate (toughened/polycarbonate)	10*	✓✓	×	×
	12†	✓✓✓	✓✓	✓
	16*	✓✓	×	×
	18†	✓✓✓	✓✓	✓
	28‡	✓✓✓	✓✓	✓✓
	46‡	✓✓✓	✓✓	✓✓
Laminate (fire)	11.4	×	×	×
	15	×	×	×
	25	×	×	×

× Not suitable for use in BS 7950, PAS 24 or LPS 1175.
✓ Suitable for use in BS 7950, PAS 24 and LPS 1175 (to SR 1).
✓✓ Suitable for use in BS 7950, PAS 24 and LPS 1175 (to SR 2).
✓✓✓ Suitable for use in BS 7950, PAS 24 and LPS 1175 (to SR 3 or above).
* Incorporating 4 mm thick polycarbonate layers.
† Incorporating 6 mm thick polycarbonate layers.
‡ Incorporating 10 mm thick polycarbonate layers.
PVB Polyvinyl butyral.

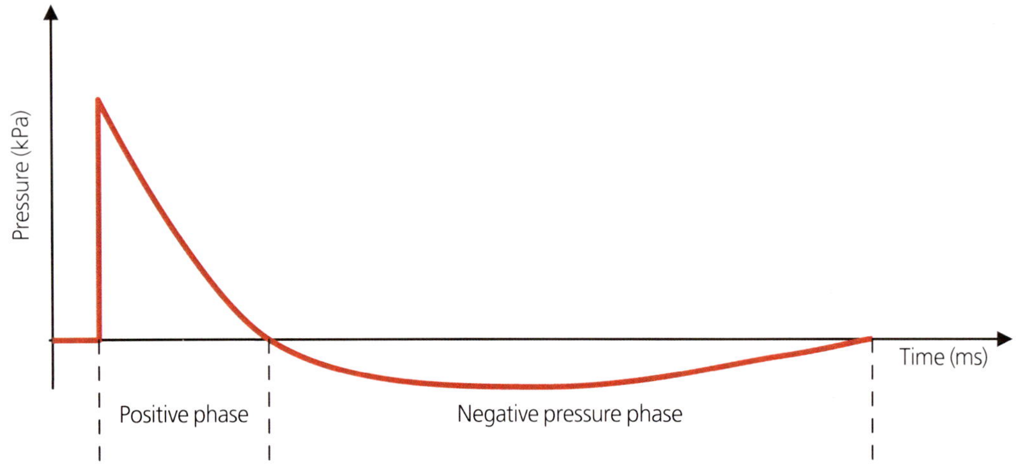

Figure 13: Typical blast wave pressure over time

The most effective way of protecting against the harmful effects of an explosion is to use laminated glass. A set of guidance notes issued by the Home Office Scientific Development Branch (HOSDB, rebranded as the Centre for Applied Science and Technology in April 2011) working in conjunction with the Centre for Protection of National Infrastructure (CPNI) laid out a number of recommendations on the thicknesses of laminated glazing to use under different circumstances[19]:

- For glazing in ordinary window frames to be a minimum thickness of 6.8 mm laminated glass fitted as either a single pane or as the internal pane of a double-glazed unit. The fitting may be strengthened (eg additional fixing screws to the beading or frame) for further slight improvements.
- For blast-resistant glazing in enhanced frames, a minimum thickness of 7.5 mm laminated glass fitted either as a single pane or as the internal pane of a double-glazed unit. Glazing to be held in by suitable gaskets in 35 mm rebates.
- Outer panes of double-glazed units to be a minimum thickness of 6 mm toughened or laminated glass.

As with impacts from criminals attempting to break through, the interlayer in laminated glass acts to retain any shards of broken glass and absorb some of the energy of the impact. The shock waves generated by the explosion force the interlayer to bulge inward, and as long as it remains intact and does not become detached from the surrounding framework the interlayer absorbs the shock waves. Furthermore, only a fine 'dust' of glass may be detached from the inner pane of the glazing, removing most of the dangers involved with flying debris.

However, some interlayer materials may be unsuitable for explosion protection. For example, acoustic resin and EVA are not as ductile as PVB and so are more likely to break up into large pieces rather than absorbing the shockwaves generated by the explosion.

The risk of injury following an explosion with different types of glazing is illustrated below (Figure 14). The hazardous area for float glass is 16 times larger than that of reduced-hazard glazing (eg laminated glass) with an enhanced framing system.

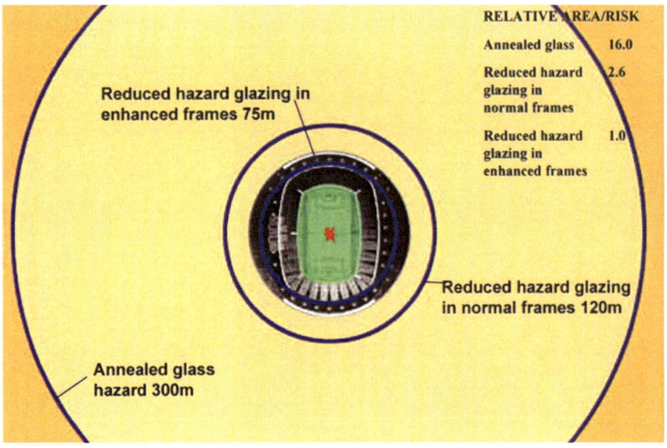

Figure 14: Hazard range for 1000 kg TNT equivalent
Reproduced by kind permission of Arup Group Ltd
Source: *Overview of blast mitigation design measures. Note 1: Structure and facades*[20]

While the distances shown in Figure 14 are indicative only, they illustrate how the choice of glass type for moderately sized windows influences the potential exposure of building occupants to glazing hazards from a large external explosion.

However, where it is not possible to use laminated glazing (for instance, in existing buildings) a few baseline measures can be put in place. These include the application of anti-shatter films (ASF) and bomb blast net curtains (BBNC).

Much like the interlayer in laminate glazing, ASF work to retain broken fragments of glass in order to prevent them from becoming hazardous debris during an explosion. They can typically be made from polyester 175 microns (0.175 mm) thick*, and should extend into the rebate to cover the entire glazing unit.

* The thicknesses stated are those typically used on buildings. However, the final choice will depend on the threat and stand-off distance. Advice should be sought from the appropriate authority.

For large areas or thick panes of glazing (eg over 6 m^2 or 8 mm thick), it is recommended by the Home Office[21] that the ASF thickness be increased to 300 microns†. The film should be applied to the inner panes of glass and should comply with the requirements of Class 1B defined in BS EN 12600 (*Glass in building – Pendulum test – Impact test method and classification for flat glass*)[22].

However, it should be noted that such films offer limited resistance to manual attack and need to be replaced periodically as some are affected by ultraviolet light over time.

BBNC are used in conjunction with ASF to further reduce the risk of glazing fragments becoming hazardous debris. However, BBNC are only effective when used in conjunction with single-glazed small-paned windows in Georgian frames. Furthermore, regular cleaning can have a detrimental effect on their performance.

BBNC must meet a number of requirements, such as being flame retardant and having a minimum mass and burst strength. They must span the full height of the window plus an additional length (which depends upon the size of the window), and twice the width of the window. More details of these requirements are laid out in the HOSDB/CPNI guidance note[19].

4.3.3 Fire resistance

Fires can cause significant damage to buildings and, in many cases, cost lives. In the UK, there were almost a million building fires and more than 4000 fire-related deaths between 2000 and 2011[23].

Fuel, oxygen and heat are the basic components of any fire, and form the familiar 'fire triangle' (Figure 15).

Although often overlooked as a fire protection measure, glazing can play a crucial role: it can act as a physical barrier to contain smoke and flames, preventing the fire from spreading from one room to another and providing vital time for people to safely evacuate.

The frame, beading, seals and other elements of the glazing system must work in conjunction with the glass itself in order to provide fire resistance for the whole glazing system. For instance, if the frame fails it may provide a small aperture for the fire to spread through, or may result in the glazing falling out of the frame entirely.

Ordinary float glass cannot withstand the extreme conditions reached during a fire, normally failing and falling out of its framing system at around 120 °C (for comparison, the flame of a candle burns at approximately 1000 °C). Fire-resistant glass is designed to survive extreme temperatures for defined periods of time (ranging from 15 to 360 minutes).

Under British and European standards, fire-resistant products are assessed against three different performance criteria, specifically loadbearing capacity (R), integrity (E) and insulation (I). Integrity and insulation are the criteria that apply to glazing.

† The thicknesses stated are those typically used on buildings. However, the final choice will depend on the threat and stand-off distance. Advice should be sought from the appropriate authority.

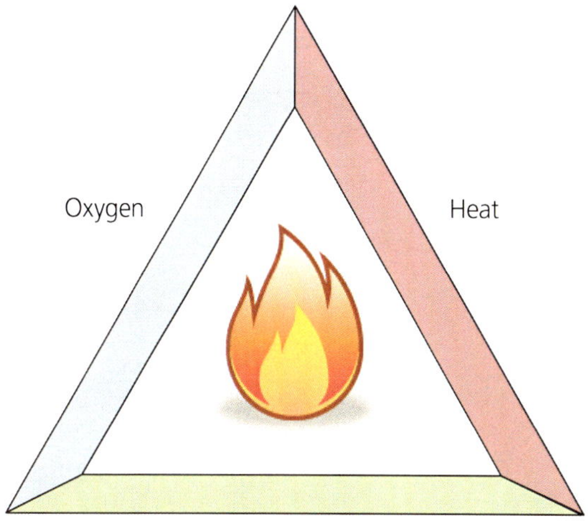

Figure 15: The fire triangle

Integrity (E) rated glass will prevent the following from occurring during a fire-resistance test for a specific duration:

- the ignition of a cotton pad held a small distance away from the glazing
- the appearance of cracks that measure more than 25 mm diameter or 6 mm wide by 150 mm long
- sustained flaming visible on the unexposed side.

Insulation (I) rated glass will prevent the following temperature rises on the unexposed side of the specimen for a specific duration:

- average temperature rise of more than 140 °C
- a temperature rise of more than 180 °C at any specific location.

Fire-resistance tests in the US include a 'hose stream test' following heating in a furnace. After heating, the glass is sprayed with water at a high pressure, replicating a sprinkler system being activated or firefighters attempting to control a blaze.

Most glass is unable to tolerate the temperature differential across its surface. With one area being hot and wanting to expand and another being cooled and contracting, the glass undergoes 'thermal shock' and shatters.

Further information on the specification and use of fire-resistant glazing systems is available in a guide produced by the Glass and Glazing Federation, *A guide to best practice in the specification and use of fire-resistant glazed systems*[24].

4.3.4 Insulation (acoustic and thermal)

Noise pollution is a serious issue, particularly when living in areas that have high levels of traffic (be that air traffic or busy roads). Stress and loss of sleep due to noise pollution can be damaging to health, and hinder learning and concentration in classrooms and workplaces. Similarly, heat loss through glazed areas can add significant sums of money to heating costs.

In a typical building facade, masonry walls will usually offer substantial levels of acoustic and thermal insulation. However, the effectiveness of the wall's insulation can be undermined by fitting the wrong type of glazing.

The level of insulation offered by glazing depends greatly upon the thickness of the panes used and, where multiple panes are installed, the pane spacing and the filling used in the cavity between them.

The pane spacing will usually determine the type of insulation afforded by the glazing. A large spacing (100 mm or more) provides the best acoustic insulation.

For thermal insulation, the optimum spacing is around 15–20 mm. If the spacing is less, there is greater heat transfer by conduction from one pane to the other. Convection becomes fully established at around 20 mm, and the thermal performance is little affected above that.

Even with an optimum spacing between panes, heat can still be transferred through the components used to keep the panes separated. These components, known as 'spacers', are typically made primarily of metal, which acts as a conductor for heat. Spacers are now increasingly being made from non-metallic materials, though some may still have a very thin metallic coating.

The cavity between panes can also affect the level of insulation. A perfect vacuum would provide the best thermal insulation but evacuated glazing, which tries to take advantage of this, is difficult to manufacture and not as commercially viable as other types of glazing at the present time, since a small leak or perforation will result in failure of the vacuum.

A more practical approach to a vacuum is to replace the air within the cavity with a chemically inert, heavy gas such as argon, krypton or xenon. These gases have lower thermal conductivity values than air, which reduces the amount of heat conducted through windows incorporating them. Other gases can be used to improve the acoustic insulation (typically only for specific frequencies), but these usually have an adverse effect on the thermal insulation.

The use of triple glazing is on the increase because it offers better levels of thermal insulation than double glazing. When the pane thicknesses and the spacing between panes vary, this also works to absorb more of the sound and increases the acoustic insulation properties.

It is still possible to achieve improved thermal insulation without using multiple panes of glass. Installing special coatings or lining the glass with films made of polyester or metal (known as 'low emissivity coatings') to reflect infrared radiation can improve its thermal insulation. Similarly, special films can be used to improve acoustic insulation.

The acoustic insulation of building elements is defined across Europe by an index known as the 'Weighted Sound Reduction Index' or R_w. The sound transmission over a range of frequencies is measured, plotted on a graph and compared with a set of reference curves. The reference curve offering the best fit against the measured curve defines the R_w, with higher values indicating better performance.

Toughened, wired and float glass of the same thickness all offer the same level of acoustic insulation. Laminated glass of the same overall thickness will usually offer a higher level of insulation than a single pane. Typical PVB interlayers will only offer very small increases in acoustic insulation, so special acoustic laminates use different types of interlayer to absorb and reduce the levels of noise pollution.

Thermal insulation can be defined in terms of its U-value, or overall heat-transfer coefficient. The U-value describes how much heat will flow through an area in a given time with a given temperature difference between the two surfaces of the glazing. Generally, the U-value of a double-glazed unit is half that of a single pane of glass because of the presence of the cavity.

Some typical values for both thermal and acoustic insulation for different generic types of glazing are offered overleaf (Table 8).

4.3.5 Safety and the Disability Discrimination Act (DDA)

Approved Document N of the Building Regulations 2010 defines a number of 'critical' locations where impacts are most likely to lead to serious cutting or piercing injuries. These likely locations are in doors and side panels, where the risk is greatest between floor and shoulder level, making hands, arms and wrists particularly vulnerable. The locations are defined as follows:

- between finished floor level and 800 mm above that level in walls
- between finished floor level and 1500 mm above that level in a door or side panel (300 mm either side of edge of door).

The regulations state that 'Glazing in critical locations should either: a) break safely, if it breaks; b) be robust or in small panes; or c) be permanently protected.'

Glazing tested to BS EN 12600 can meet these requirements, and will be given a three-figure classification that describes if and how the glazing breaks. The figures describe the drop heights at which the specimens either did not break or broke safely, and the mode of breakage (either sharp fragments as in float glass, 'spider web' cracks as in laminated glass, or disintegration as in toughened glass).

Glazing tested to LPS 1270 is formed of laminates to provide adequate levels of resistance to manual attack. As such, it generally conforms to the safety standards of BS EN 12600. However, it is important to ensure that double- and triple-glazing units formed from a mixture of panes that meet LPS 1270, and others that do not, satisfy the requirements of Approved Document N. This is because the panes that do not meet LPS 1270 may not break safely.

Approved Document N and Approved Document M: Access to and use of buildings[25] advise on the subject of 'manifestation' of glazing for safety. This applies where the presence of large, uninterrupted areas of transparent glazing may not be obvious to people, including those with visual impairment, who therefore may collide with it. For instance, the presence of a fully-glazed wall without transoms or mullions between the panes of glass in a reception area may not immediately be obvious, which could cause a hazard.

Table 8: Typical acoustic and thermal insulation values for generic types of glazing

Glazing type	Thickness (mm)	Insulation ratings	
		Acoustic (dB R_w)	Thermal U-value (W/m^2 K)
Float/wired/toughened glass (single glazing)	4	29	5.6
	6	31	5.6
	10	33	5.6
Float/wired/toughened glass (double glazing)	4/16/4	29	~ 3
	4/16/6	32	~ 3
	4/16/10	35	~ 3
Low emissivity (double glazing)	4/16/4	Not known	~ 2
Laminate (float/PVB)	6.4	32	5.7
Laminate (float/PVB)	11.4	Not known	5.7
Laminate (float/PVB)	11.5	34	5.6
Laminate (toughened/polycarbonate)	16	Not known	5.33
Laminate (toughened/polycarbonate)	18	Not known	5.09
Laminate (fire)	15	38	Not known

PVB Polyvinyl butyral.

To make the glazing obvious, decorations such as broken lines, solid lines or company logos should be applied to the surface of the glazing at two levels: 850 mm to 1000 mm, and 1400 mm to 1600 mm above floor level. Logos or signs must be 150 mm high, whereas decorative lines or bands must be at least 50 mm high. Additionally, high-contrast strips should be placed around the edges of glazed entrance doors adjacent to glazed screens in order to differentiate the two, and glazed doors should be protected by guarding down the leading edge to prevent hazards.

Increasingly, phosphorescent materials are being applied to produce patterns that make the surface of the glazing obvious during daytime, and can provide a self-illuminating pattern during hours of darkness.

4.4 Comparison of glazing types

A brief overview of the different performance characteristics offered by various generic types of glazing is given in Table 9. It is provided only as an illustrative guide, and any performance declared by manufacturers should be confirmed by independent verification wherever possible.

With reference to Table 9, it should be noted that the column headed 'fire performance' includes both 'reaction to fire' and 'fire resistance' classifications. These are two very distinct fire performance classifications, characteristic of different stages of a fire. 'Reaction to fire' is an evaluation of the contribution of a construction product to the initial stages of fire growth and 'fire resistance integrity' is an evaluation of the construction product's ability to contain a fully developed fire.

Table 9: Comparison of generic glazing properties

Glazing type	Thickness (mm)	Mass (kg/m²)	Relative cost*	Acoustic rating (dB R_w)	Fire performance	Safety rating to BS EN 12600:2002	Security rating to BS EN 356:2000	Thermal U-value (W/m² K)
Toughened glass	4	10	£	29	–	1C3	–	5.6
Polycarbonate	6	7.3	££	–	1Y†	–	P5A	5.09
Wired glass	6	15	££	31	E60‡	–	–	5.6
Laminate (float/PVB)	6.8	16	£	32	–	1B1	P1A	5.7
Float glass	10	25	£	33	–	–	–	5.6
Polycarbonate	10	12	£££	–	1Y†	–	–	4.61
Toughened glass	10	25	££	33	–	1C1	–	5.6
Laminate (float/PVB)	11.5	27	££	34	–	1B1	P5A	5.6
Laminate (toughened/polycarbonate)	12	20	£££	–	–	–	P7B	5.09
Laminate (fire)	15	36	££££	38	EI30‡	1B1	–	–
Laminate (fire)	23	55	££££	40	EI60‡	1B1	–	–
Laminate (toughened/polycarbonate)	46	63	£££££	–	–	–	P8B	4.61

* Relative cost is represented on a scale from £ (least expensive) to £££££ (most expensive).

† Class 1Y in accordance with BS 476-7 (*Fire tests on building materials and structures – Part 7: Method of test to determine the classification of the surface spread of flame of products*)[26]. The standard classifies products from 1 (best performance) to 4 (worst performance). The 'Y' suffix indicates that softening and/or other behaviour has occurred during testing that was considered to have affected the result.

‡ Classification in accordance with BS EN 13501-2:2007 + A1:2009 (*Fire classification of construction products and building elements – Part 2: Classification using data from fire resistance tests, excluding ventilation services*)[27].

– Performance not known or undeclared.

E Integrity.

I Insulation.

PVB Polyvinyl butyral.

5 Conclusions

In recent times, increased design complexity and specific emphasis on the environmental and energy performance of today's buildings have meant that new glazing systems and materials have been developed and are in use.

However, burglary statistics for England and Wales in 2010–2011 show that approximately 25% of incidents involve breaking glass. This is thought to be due to an improvement in other hardware standards and specifications shifting the focus of attacks to the glass, which itself had remained relatively unchanged.

Glass and glass laminates can be designed and specified for a range of different security and environmental requirements. Correct specification will depend on a number of product performance factors.

BRE Trust funded a research project to investigate the performance of glazing in security applications. The extensive test programme conducted found that, generally, glass and glazing systems that have not been laminated with plastic material show very low resistance to physical attack in relation to the building products and materials that the glass may be fitted into.

The tests also identified the fact that laminated glazing incorporating layers of polycarbonate offered the greatest resistance to manual attack, and that specifying glazing by thickness alone was not sufficient to provide adequate levels of resistance.

It is important to consider the threat, location and general constraints of the glazing before selection, and a specification sequence has been developed to assist this process (Figure 9). The aim is to try to ensure that the glazing selected will offer an appropriate level of resistance to physical attack by criminals, and use of this sequence will ensure that additional features, such as thermal and acoustic insulation, have been taken into consideration.

Although different types of glazing may be vulnerable to the creation of an aperture large enough for a criminal to climb through, glazing may be even more vulnerable to the creation of smaller apertures. This is illustrated in Table 5 and Table 6. Although not large enough to climb through, a smaller aperture may permit a criminal to pass a variety of tools or reach through it to operate internal hardware fitted to windows and doors, or remove any valuable items within reach of the aperture.

Because windows and doors may be fitted with a range of deadlocked and non-deadlocked hardware, specifiers may find Table 7 a useful reference for determining whether the type of glazing selected may be suitable for use with that particular type of hardware. The choice of glazing should always be supported by relevant test data to confirm whether or not it is suitable.

If there is any doubt of a glazing product's resistance to manual attack, it is recommended that the glazing be approved by an independent third party to a relevant performance standard, such as LPS 1270 (see www.redbooklive.com).

6 References and further information

6.1 References

1. Home Office. British crime survey 2010/11. London, Home Office, 2011.
2. Home Office. Crime in England and Wales 2010/11: supplementary tables: nature of burglary, vehicle-related theft, bicycle theft, other household theft, personal and other theft, vandalism and violent crime. London, Home Office, 2011.
3. City of London Crime and Disorder Reduction Partnership (CDRP). The City of London Crime and Disorder Reduction Partnership 2005 audit report. London, CDRP, 2005.
4. Pascoe T and Harrington-Lynn J. The role of windows in domestic burglary. BRE IP 18/94. Bracknell, IHS BRE Press, 1994.
5. BSI. Specification for enhanced security performance of windows for domestic applications. BS 7950:1997 (incorporating Amendments 1, 2 and 3). London, BSI, 2007.
6. BSI. Enhanced security performance requirements for door assemblies – Part 1: Single and double leaf, hinged external door assemblies to dwelling. PAS 24:2007. London, BSI, 2007.
7. Loss Prevention Certification Board (LPCB). Requirements and testing procedures for the LPCB approval and listing of intruder-resistant building components, strongpoints, security enclosures and free-standing barriers. LPS 1175: Issue 7. Watford, BRE Global, 2010.
8. US Department of State. Certification standard – Forced entry and ballistic resistance of structural systems. SD-STD-01.01, Revision G (amended). Washington, DC, US Department of State, 1993.
9. Loss Prevention Certification Board (LPCB). Requirements and testing procedures for the LPCB approval and listing of intruder-resistant security glazing units. LPS 1270: Issue 1. Watford, BRE Global, 2010.
10. BSI. Glass in building – Security glazing – Testing and classification of resistance against manual attack. BS EN 356:2000. London, BSI, 2000.
11. Department for Communities and Local Government (DCLG). The Building Regulations 2010. Approved Document N: Glazing – safety in relation to impact, opening and cleaning (1998 edn incorporating 2000 and 2010 amendments). London, DCLG, 2010.
12. BSI. Glass in building – Basic soda lime silicate glass products – Part 2: Float glass. BS EN 572-2:2004. London, BSI, 2004.
13. BSI. Glass in building – Insulating glass units – Part 1: Generalities, dimensional tolerances and rules for the system description. BS EN 1279-1:2004. London, BSI, 2004.
14. BSI. Glass in building – Laminated glass and laminated safety glass. BS EN ISO 12543-1:1998. London, BSI, 1998.
15. BSI. Glass in building – Thermally toughened soda lime silicate safety glass. BS EN 12150-1:2000. London, BSI, 2000.
16. BSI. Pedestrian doorsets, windows, curtain walling, grilles and shutters – Burglar resistance – Requirements and classification. BS EN 1627:2011. London, BSI, 2011.
17. BSI. Glass in building – Security glazing – Testing and classification of resistance against bullet attack. BS EN 1063:2000. London, BSI, 2000.
18. BSI. Glass in building – Security glazing – Testing and classification of resistance against explosion pressure. BS EN 13541:2001. London, BSI, 2001.
19. Home Office. Guidance Note: Glazing enhancement to improve blast resistance. HOSDB 11/08. London, Home Office, 2008.
20. Hadden D. Overview of blast mitigation design measures. Note 1: Structure and facades. London, Arup Security Consulting, 2003.
21. Home Office. Guidance Note: Use of anti-shatter film (ASF) and bomb blast net curtains (BBNC). HOSDB 11A/08. London, Home Office, 2008.
22. BSI. Glass in building – Pendulum test – Impact test method and classification for flat glass (incorporating corrigendum April 2010). BS EN 12600:2002. London, BSI, 2002.
23. Department for Communities and Local Government (DCLG). Fire statistics: Great Britain, 2010–2011, tables 1a–16. London, DCLG, 2011.
24. Glass and Glazing Federation. A guide to best practice in the specification and use of fire-resistant glazed systems. London, Glass and Glazing Federation, 2009.
25. Department for Communities and Local Government (DCLG). The Building Regulations 2010. Approved Document M: Access to and use of buildings (2004 edn incorporating 2010 amendments). London, DCLG, 2010.
26. BSI. Fire tests on building materials and structures – Part 7: Method of test to determine the classification of the surface spread of flame of products. BS 476-7:1997. London, BSI.1997.
27. BSI. Fire classification of construction products and building elements – Part 2: Classification using data from fire resistance tests, excluding ventilation services. BS EN 13501-2:2007 + A1:2009. London, BSI, 2009.

6.2 Other sources of guidance

1. Centre for the Protection of National Infrastructure, www.cpni.gov.uk
2. Centre for Applied Science and Technology, www.homeoffice.gov.uk/science-research/cast
3. Association of Chief Police Officers, Secured by Design, www.securedbydesign.com
4. Association of British Insurers, www.abi.org.uk

Appendix

A.1 Supplementary data on existing standards

Tables A1–A4 contain information on the classifications within different standards. Table A5 contains details of a number of other existing domestic and international standards that tackle glazed building products.

Table A1: Security rating classifications in LPS 1175[7]

Security rating	Working time (minutes)	Tool category	Typical tools
SR 1	1	A	Small knife, small screwdriver, small lever, spanners, tweezers, wire, hooks, wooden or plastic wedges
SR 2	3	B	Category A tools plus claw hammer, hand drill, junior hacksaw, larger screwdrivers and bolt cutters
SR 3	5	C	Category A and B tools plus hand axe, club hammer, cold chisels, 7.2 V drill, hacksaw and scissor jack
SR 4	10	D	Category A, B and C tools plus 12 V battery-powered tools, larger bolt cutters, hooligan bar, sledgehammer and steel wedges
SR 5	20	D+	Category A, B and C tools plus 18 V battery-powered tools, larger bolt cutters, hooligan bar, sledgehammer and steel wedges
SR 6	20	E	Category A, B, C and D+ tools plus up to 1100 W mains-powered power tools, hole saw and pinch bar
SR 7	20	F	Category A, B, C, D+ and E tools plus up to 2300 W mains-powered power tools and oxyacetylene cutting kit
SR 8	30	G	Category A, B, C, D+, E and F tools plus chainsaw, 5 kW cut-off saw, hydraulic head and toe jack and pneumatic impact tool

Table A2: Classifications in BS EN 356[10]

Classification	Test method	Drop height (mm)	Number of strikes
P1A	Hard body drop test	1500	3 in a triangle
P2A	Hard body drop test	3000	3 in a triangle
P3A	Hard body drop test	6000	3 in a triangle
P4A	Hard body drop test	9000	3 in a triangle
P5A	Hard body drop test	9000	3 × 3 in a triangle
P6B	Hammer and axe strikes	–	30–50
P7B	Hammer and axe strikes	–	51–70
P8B	Hammer and axe strikes	–	Over 70

Table A3: Classifications in BS EN 1063[17]

Classification	Weapon	Calibre	Number of strikes
BR1	Weapon	0.22 LR	3
BR2	Rifle	9 mm Luger	3
BR3	Hand gun	0.357 Magnum	3
BR4	Hand gun	0.44 rem. Magnum	3
BR5	Rifle	5.56 × 45	3
BR6	Rifle	7.62 × 51	3
BR7	Rifle	7.62 × 51	3
SG1	Shotgun	Cal. 12/70	1
SG2	Shotgun	Cal. 12/70	3

Table A4: Classifications in BS EN 13541[18]

Classification	Positive maximum overpressure of reflected blast wave Pr (kPa)	Duration of positive pressure phase (ms)
ER1	$50 \leq Pr < 100$	≥ 20
ER2	$100 \leq Pr < 150$	≥ 20
ER3	$150 \leq Pr < 200$	≥ 20
ER4	$200 \leq Pr < 250$	≥ 20

Table A5: Summary of existing glazing standards

Standard	Title	Notes
ASTM F 476-84 (reapproved 2002)	Standard test methods for security of swing door assemblies	Specifies a 200 J impact test on glazing near to the locking hardware
ASTM F 1233-98 (reapproved 2004)	Standard test method for security glazing materials and systems	Specifies manual attacks on glazing using blunt and blunt impact tools, and tools to induce thermal stress and chemical deterioration
BS 644:2003	Timber windows – Factory assembled windows of various types – Specification	Recommends windows meet the requirements of BS 7950*
BS 4873:2004	Aluminium alloy windows	Recommends windows meet the requirements of BS 7950*
BS 5544:1978 (incorporating Amendment 1)	Specification for anti-bandit glazing (glazing resistant to manual attack)	Specifies a number of successive low-energy impacts followed by high-energy impacts and a boil test for laminated glass
BS 7950:1997 (incorporating Amendments 1, 2 and 3)	Specification for enhanced security performance of windows for domestic applications	Specifies that glazing used on windows with non-key-locking hardware on the inside should be laminated (but does not specify thickness). Tests are conducted on the glazing retention, but not on the glazing itself
BS EN 356:2000	Glass in building – Security glazing – Testing and classification of resistance against manual attack (Europe)	Current EU standard for classifying glazing resistance to manual attack, which replaced BS 5544. This standard specifies mechanical impact tests rather than manual attacks
BS EN 1627:2011	Windows, doors, shutters – Burglar resistance – Requirements and classification	Specifies the glazing must meet BS EN 356
ISO 16936	Glass in building – Forced-entry security glazing	Split into four parts, specifying the following tests: • repetitive ball drop • hammer and axe impact • manual attack • pendulum impact (under thermal and fire stress)

* BS 7950 does not specify any performance of the glazing itself.

Table A5 continued: Summary of existing glazing standards

Standard	Title	Notes
LPS 1175: Issue 7	Requirements and testing procedures for the LPCB approval and listing of intruder-resistant building components, strongpoints, security enclosures and free-standing barriers	Glazing is treated as an integral component of the building element tested and therefore requires the glazing to offer compatible levels of resistance to manual attack
LPS 1270: Issue 1	Requirements and testing procedures for the LPCB approval and listing of intruder-resistant security glazing units	Building upon the principles of LPS 1175, this standard defines the resistance to creation of various sized apertures based on hole sizes typically required to operate non-deadlocked hardware through (such as panic exit devices or thumb turns) or enter through fixed lights or units with key deadlocking on both sides
PAS 24:2007 + A2:2011	Enhanced security performance requirements for door assemblies – Part 1: Single and double leaf, hinged external door assemblies to dwellings	Specifies that glazing used on doorsets with non-key-locking hardware on the inside should be laminated (but does not specify thickness). Specifies that side/fan lights meet the requirements of BS 7950*
SD-STD-01.01, Revision G (amended 30/04/1993)	Certification standard – Forced entry and ballistic resistance of structural systems (US Department of State, USA)	Glazing is treated as an integral component of the building element. The glazing is therefore expected to offer the same level of resistance to manual attack
ULC-S332-93	Standard for burglary-resisting glazing material (Independent, Canada)	Conducts a series of low-energy impacts, thermal conditioning and high-energy impacts

* BS 7950 does not specify any performance of the glazing itself.

A.2 British Crime Survey data

Table A6: British Crime Survey results 2002–2011

Method of entry	Percentage of all burglary crimes (burglary with entry and attempted burglary)								
	2002/03	2003/04	2004/05	2005/06	2006/07	2007/08	2008/09	2009/20	2010/11
Door	**63**	**63**	**68**	**69**	**70**	**70**	**72**	**73**	**72**
Forced lock	27	26	26	25	25	28	31	29	27
Door was not locked	14	15	17	21	16	15	17	15	15
Broke or cut panel of door	10	10	11	9	9	10	10	10	10
Pushed last person to enter	4	5	6	5	8	6	4	9	7
They had a key	1	4	4	6	5	5	5	4	5
False pretences	4	5	5	5	3	4	4	4	5
Kicked, smashed or rammed door	–	–	–	–	2	3	1	2	1
Broke or smashed glass of door	–	–	–	–	1	1	1	1	0
Other method	7	4	3	3	5	4	4	4	6
Window	**36**	**33**	**32**	**28**	**29**	**29**	**26**	**24**	**26**
Forced lock or catch	16	15	15	12	13	12	9	10	11
Broke or cut glass	12	10	10	9	8	9	9	9	9
Window open or not secured	8	8	7	7	8	8	7	4	6
Other method	0	1	1	1	1	0	1	2	1
Other	**2**	**2**	**2**	**3**	**3**	**2**	**1**	**1**	**2**

Source: *Crime in England and Wales 2010/11: supplementary tables: nature of burglary* (Table 1.3)[2].
Note: Percentages sum to more than 100 as more than one response was allowed.

Other reports from BRE Trust

Subsidence damage to domestic buildings: lessons learned and questions remaining. **FB 1**

Potential implications of climate change in the built environment. **FB 2**

Behaviour of concrete repair patches under propped and unpropped conditions: critical review of current knowledge and practices. **FB 3**

Construction site security and safety: the forgotten costs! **FB 4**

New fire design method for steel frames with composite floor slabs. **FB 5**

Lessons from UK PFI and real estate partnerships: drivers, barriers and critical success factors. **FB 6**

An audit of UK social housing innovation. **FB 7**

Effective use of fibre reinforced polymer materials in construction. **FB 8**

Summertime solar performance of windows with shading devices. **FB 9**

Putting a price on sustainability. BRE Centre for Sustainable Construction and Cyril Sweett. **FB 10**

Modern methods of house construction: a surveyor's guide. **FB 11**

Crime opportunity profiling of streets (COPS): a quick crime analysis– rapid implementation approach. **FB 12**

Subsidence damage to domestic buildings: a guide to good technical practice. **FB 13**

Sustainable refurbishment of Victorian housing: guidance, assessment method and case studies. **FB 14**

Putting a price on sustainable schools. **FB 15**

Knock it down or do it up? **FB 16**

Micro-wind turbines in urban environments: an assessment. **FB 17**

Siting micro-wind turbines on house roofs. **FB 18**

Automatic fire sprinkler systems: a guide to good practice. **FB 19**

Complying with the Code for Sustainable Homes: lessons learnt on the BRE Innovation Park. **FB 20**

The move to low-carbon design: are designers taking the needs of building users into account? **FB 21**

Building-mounted micro-wind turbines on high-rise and commercial buildings. **FB 22**

The real cost of poor housing. **FB 23**

A guide to the Simplified Building Energy Model (SBEM): what it does and how it works. **FB 24**

Vacant dwellings in England: the challenges and costs of bringing them back into use. **FB 25**

Energy efficiency in new and existing buildings: comparative costs and CO_2 savings. **FB 26**

Health and productivity benefits of sustainable schools: a review. **FB 27**

Integrating BREEAM throughout the design process: a guide to achieving higher BREEAM and Code for Sustainable Homes ratings through incorporation with the RIBA Outline Plan of Work and other procurement routes. **FB 28**

Design fires for use in fire safety engineering. **FB 29**

Ventilation for healthy buildings: reducing the impact of urban pollution. **FB 30**

Financing UK carbon reduction projects. **FB 31**

The cost of poor housing in Wales. **FB 32**

Dynamic comfort criteria for structures: a review of UK standards, codes and advisory documents. **FB 33**

Water mist fire protection in offices: experimental testing and development of a test protocol. **FB 34**

Airtightness in commercial and public buildings. 3rd edn. **FB 35**

Biomass energy. **FB 36**

Environmental impact of insulation. **FB 37**

Environmental impact of vertical cladding. **FB 38**

Environmental impact of floor finishes: incorporating The Green Guide ratings for floor finishes. **FB 39**

LED lighting. **FB 40**

Radon in the workplace. 2nd edn. **FB 41**

U-value conventions in practice. **FB 42**

Lessons learned from community-based microgeneration projects: the impact of renewable energy capital grant schemes. **FB 43**

Energy management in the built environment: a review of best practice. **FB 44**

The cost of poor housing in Northern Ireland. **FB 45**

Ninety years of housing, 1921–2011. **FB 46**

BREEAM and the Code for Sustainable Homes on the London 2012 Olympic Park. **FB 47**

Saving money, resources and carbon through SMARTWaste. **FB 48**

Concrete usage in the London 2012 Olympic Park and the Olympic and Paralympic Village and its embodied carbon content. **FB 49**

A guide to the use of urban timber. **FB 50**

Low flow water fittings: will people accept them? **FB 51**

Evacuating vulnerable and dependent people from buildings in an emergency. **FB 52**

Refurbishing stairs in dwellings to reduce the risk of falls and injuries. **FB 53**

Dealing with difficult demolition wastes. **FB 54**